The Unwritten
LAWS OF ENGINEERING

with revisions and additions
by

James G. Skakoon

and original
by

W. J. King

ASME PRESS • NEW YORK • 2001

PREFACE

When my editor inquired about my interest in updating *The Unwritten Laws of Engineering*, I was delighted for the opportunity. The book had first been published as three articles in *Mechanical Engineering* in 1944, and was reprinted most recently in 1994, substantially unchanged from the original. When first reading it a few years ago, I was surprised at how well its advice had held up over the decades. Except for a few omissions and outdated laws, it seemed equally as fitting then as it must have been over fifty years before, and it was a lot of fun to read.

The biggest challenge was updating the book without ruining it altogether. Indeed, a common response from readers on what to change was: "Don't change anything. Why would you want to?" Well, some things just plain demanded it. At the same time, a new edition needed to retain timelessness; after all, the original survived more than half a century in fairly good health.

So, this updated edition keeps the style of the original and much of its content. Changes have to do with shifted societal values, changed employment laws, and evolved corporate structures. No apology is offered for the lack of reference to the latest technology (e-mail, computers, internet) this book's advice transcends the mere implements of the engineer. Some words and phrases in the original were painfully archaic and have been changed, but many of the old-fashioned words remain — they add to the fun.

Although I wish to thank all those who helped with the text, Jennifer Kern, Dr. Steven Kern, and John Gillespie get special, personal thank-you's. Above all, I wish to acknowledge the original author, W. J. King. He mined and cut a little gem, and I am privileged to have been able to reset it.

James G. Skakoon

CONTENTS

INTRODUCTION

Prior to writing the first text of this book, the originating author admitted to having become very much aware, which can be observed in any engineering organization, that the chief obstacles to the success of individual engineers or of groups of engineers are of a personal and administrative rather than a technical nature. It was apparent that both he and his associates were getting into much more trouble by violating the undocumented laws of professional conduct than by committing technical sins against the well-documented laws of science. Since the former appeared to be unwritten at that time, "laws" were formulated and collected into a scrapbook as a professional code of sorts. Although they were, and in this latest edition still are fragmentary and incomplete, they are offered here for whatever they may be worth to younger engineers just starting their careers, and to older ones who know these things perfectly well but who all too often fail to apply them.

None of these laws is theoretical or imaginary, and however obvious they may appear, their repeated violation is responsible for much of the frustration and embarrassment to which engineers everywhere are liable. In fact, the first edition of this book was primarily a record derived from direct observation over 17 years of four engineering departments, three of them newly organized and struggling to establish themselves by trial-and-error. It has been supplemented, confirmed, and updated by the experience of others as gathered from numerous discussions, observations, and literature, so that it most emphatically does not reflect the unique experience or characteristics of any one organization.

Many of these laws are generalizations to which exceptions will occur in special circumstances. There is no thought of urging a servile adherence to rules and red tape, for there is no substitute for judgment; vigorous individual initiative is needed to cut through formalities in emergencies. But in many respects these laws are like the basic laws of society; they cannot be violated too often with impunity, notwithstanding striking exceptions in individual cases.

PART I

WHAT THE BEGINNER NEEDS TO LEARN AT ONCE

IN RELATION TO THE WORK

However menial and trivial your early assignments may appear, give them your best efforts.

Many young engineers feel that the minor chores of a technical project are beneath their dignity and unworthy of their college training. They expect to prove their true worth in some major, vital enterprise. Actually, the spirit and effectiveness with which you tackle your first humble tasks will very likely be carefully watched and may affect your entire career.

Occasionally you may worry unduly about where your job is going to get you — whether it is sufficiently strategic or significant. Of course these are pertinent considerations and you would do well to take some stock of them. But by and large, it is fundamentally true that if you take care of your present job well, the future will take care of itself. This is particularly so within large corporations, which constantly search for competent people to move into more responsible positions. Success depends so largely upon personality, native ability, and vigorous, intelligent prosecution of any job that it is no exaggeration to say that your ultimate chances are much better if you do a good job on some minor detail than if you do a mediocre job as a project leader. Furthermore, it is also true that if you do not first make a good showing on your present job you are not likely to be given the opportunity to try something else more to your liking.

Demonstrate the ability to get things done.

This is a quality that may be achieved by various means under different circumstances. Specific aspects will be elaborated in some of the succeeding paragraphs. It can probably be reduced, however, to a combination of three basic characteristics:

- initiative, which is expressed in energy to start things and aggressiveness to keep them moving briskly,
- resourcefulness or ingenuity, i.e., the faculty for finding ways to accomplish the desired result, and

- persistence (tenacity), which is the disposition to persevere in spite of difficulties, discouragement, or indifference.

This last quality is sometimes lacking in the make-up of brilliant engineers to such an extent that their effectiveness is greatly reduced. Such dilettantes are known as "good starters but poor finishers." Or else it will be said: "You can't take their type too seriously; they will be all steamed up over an idea today, but by tomorrow will have dropped it for some other wild notion." Bear in mind, therefore, that it may be worthwhile finishing a job, if it has any merit, just for the sake of finishing it.

In carrying out a project, do not wait passively for anyone — suppliers, sales people, colleagues, supervisors — to make good on their delivery promises; go after them and keep relentlessly after them.

This is one of the first things a new engineer must learn in entering a manufacturing organization. Many novices assume that it is sufficient to make a request or order, then sit back and wait until the goods or services are delivered. Most jobs progress in direct proportion to the amount of follow-up and *expediting* that is applied to them. Expediting means planning, investigating, promoting, and facilitating every step in the process. Cultivate the habit of looking immediately for some way around each obstacle encountered, some other recourse or expedient to keep the job rolling without losing momentum.

On the other hand, the matter is occasionally overdone by overzealous individuals who make themselves obnoxious and antagonize everyone with their incessant pestering. Be careful about demanding action from others. Too much insistence and agitation may result in more damage to one's personal interest than could ever result from the miscarriage of the item involved.

Confirm your instructions and the other person's commitments in writing.

Do not assume that the job will be done or the bargain kept just because someone agreed to do it. Many people have poor memories, others are too busy, and almost everyone will take the matter a great deal more seriously if it is in writing. Of course there are exceptions, but at times it pays to copy a third person as a witness.

When sent out on a business trip of any kind, prepare for it, execute the business to completion, and follow up after you return.

Any trip to the field, whether for having a design review, resolving a complaint, analyzing a production problem, investigating a failure, calling on a customer, visiting a supplier, or attending a trade show, deserves your special attention to return the maximum benefit for the time and expense. Although each business trip will be unique, and the extent to which you must do the following will be different for each, as a minimum, be sure to:

- *Plan the travel.* This is more than just reserving transportation and hotels. Consider all eventualities such as lost luggage, missed connections, late arrivals, unusual traffic. Those you are meeting have arranged their schedules for you, so don't disappoint them — arrive on time and ready to perform. Follow the motto: "If you can't be on time, be early!"
- *Plan and prepare for the business to be done.* Prepare and distribute agendas before you arrive. Send ahead any material to be reviewed. Be sure everything (e.g., samples, prototypes, presentations) is complete. Practice any presentations, however minor they might seem, beforehand. In short, be fully prepared and allow those you visit to prepare fully.
- *Complete the business at hand.* You will not always be able to carry out a business trip to your complete satisfaction; others may control the outcome to a different conclusion. Nevertheless, if you have been sent out to complete a specific task, perhaps to analyze a failure or observe a product in use, and the allotted time proves inadequate for whatever reason, stay until the job is complete. Neither your supervisor nor those you visit will like it if another engineer has to be sent out later to finish what you did not.

- *Execute the appropriate follow-up.* Often a seemingly successful trip will come to nothing without adequate follow-up. Use meeting minutes, trip reports, and further communications to your best advantage.

Develop a "let's go see!" attitude.

Throughout your career people will approach you with all manner of real-life problems they will have observed on devices or equipment for which you have responsibility. A wonderfully effective response, both technically and administratively, is to invite them to have a look with you — i.e., "Let's go see!" It is seldom adequate to remain at one's desk and speculate about causes and solutions, or to retreat to drawings, specifications, and reports and hope to sort it all out. Before ever being able to solve a problem, you will need abundant insight, insight that can only be developed by observing first-hand what might be at once too subtle and complex only to imagine (Ferguson, p. 56).

Avoid the very appearance of vacillating.

One of the gravest personal indictments is to have it said that an engineer's opinion at any time depends merely upon the last person with whom he or she has spoken. Refrain from stating an opinion or promoting an undertaking until you have had a reasonable opportunity to obtain and study the facts. Thereafter see it through if at all possible, unless fresh evidence makes it folly to persist. Obviously the extremes of obstinacy and dogmatism should be avoided, but remember that reversed decisions could be held against you.

Don't be timid — speak up — express yourself and promote your ideas.

Too many new employees seem to think that their job is simply to do what they are told. Of course there are times when it is wise and prudent to keep silent, but, as a rule, it pays to express your point of view whenever you can contribute something. The quiet, timorous individual who says nothing is usually credited with having nothing to say.

It frequently happens in any sort of undertaking that nobody is sure of just how a matter ought to be handled; it's a question of selecting some kind of program with a reasonable chance of success. This is commonly to be observed in design or project meetings. The first person to speak up with a definite and plausible proposal often has a better-than-even chance of carrying the floor, provided only that the scheme is definite and plausible. (The "best" scheme usually cannot be recognized as such in advance.) It also happens that the one who talks most knowingly and confidently about the project will often be assigned to carry it out. If you do not want the job, say nothing and you'll be overlooked, but you'll also be overlooked when it comes time to assign larger responsibilities.

Strive for conciseness and clarity in oral or written reports.

If there is one most irksome encumbrance to promoting urgency in the workplace, it is the person who takes a half-hour of rambling discourse to say what could be said in one sentence of 20 words. There is a curious and widespread tendency among engineers to surround the answer to a simple question with so many preliminaries and commentaries that the answer itself can hardly be discerned. It is so difficult to get a direct answer out of some engineers that their usefulness is thereby greatly diminished. The tendency is to explain the answer before answering the question. To be sure, very few questions endure simple answers without qualifications, but the important thing is to state the essence of the matter as succinctly as possible first. On the other hand, there are times when it is important to add the pertinent background or other relevant facts to illuminate a simple statement. The trick is to convey the maximum of significant information in the minimum time, a valuable asset to anyone.

An excellent guide in this respect may be found in the literary construction called the "inverted pyramid." Start at the bottom —

the beginning — with the single most important fact, the one the audience must know before learning more. Often this is the conclusion itself. Progressively broaden the pyramid by constructing each sentence to build upon its predecessor. In this way you will be able to clearly explain even complicated, abstract concepts to anyone. Even if by the end the explanation has become too complex for some, you can take smug comfort knowing that, because you began with your primary point, or the conclusion, or the simple answer, everyone understands you. You can hardly do better than to adopt this method in your communication, presenting your facts in the order of importance, as journalists often do, as if you might be cut off at any minute.

Be extremely careful of the accuracy of your statements.

This seems almost trite, and yet many engineers lose the confidence of their superiors and associates by habitually guessing when they do not know the answer to a direct question. It is important to be able to answer questions concerning your responsibilities, but a wrong answer is worse than no answer. If you do not know, say so, but also say, "I'll find out right away." If you are not certain, indicate the exact degree of certainty or approximation upon which your answer is based. A reputation for dependability and reliability can be one of your most valuable assets.

This applies, of course, to written matter, calculations, etc., as well as to oral reports and discussions. It is bad business to submit a report for approval without first carefully checking it yourself, and yet formal reports are sometimes turned in, or worse, sent out, full of glaring errors and omissions.

IN RELATION TO YOUR SUPERVISOR

Every manager must know what goes on in his or her domain.

This principle is so elementary and fundamental as to be axiomatic. It follows very obviously that a manager cannot possibly manage a department successfully without knowing what's going on in it. This applies as well to project managers with specific responsibilities but without direct subordinates as it does to department heads. No sensible person will deny the soundness of this principle and yet it is commonly violated or overlooked. It is cited here because several of the rules that follow are concerned with specific violations of this cardinal requirement.

One of the first things you owe your supervisor is to keep him or her informed of all significant developments.

This is a corollary of the preceding rule: "Every manager must know what goes on. . . ." The main question is: How much must a manager know — how many of the details? This is always a difficult matter for the new employee to get straight. Many novices hesitate to bother their superiors with everyday minutiae, and it is certainly true that it can be overdone in this direction, but in by far the majority of cases the manager's problem is to extract enough information to keep adequately posted. It is a much safer course to risk having your supervisor say, "Don't bother me with so many details," than to allow your supervisor to say, "Why doesn't someone tell me these things?" Bear in mind that your manager is constantly called upon to account for, defend, and explain your activities to others, as well as to coordinate these activities into a larger plan. Compel yourself to provide all the information that is needed for these purposes.

No matter how hard you try nor how good an engineer you become, technical difficulties — unexpected problems or failures — will occur that you will dread having to inform your supervisor about. The best you can hope to do is to develop solutions to such problems so that you can present these along with the problem, and so that they can be implemented with the greatest urgency. No manager will like being surprised by unanticipated problems (although you are obligated to report them without hesitation), but you will improve your predicament immeasurably if you also bring solid recommendations for solutions.

Do not overlook the steadfast truth that your direct supervisor is your "boss."

This sounds simple enough, but some engineers never get it. By all means, you are working for society, the company, the department, your project team, your project leader, your family, and yourself — but primarily you should be working for and through your supervisor, the manager to whom you directly report.

You will no doubt encounter conflicts — you are assigned to a project team with a demanding leader, a corporate executive orders a task be done, and so forth. When this happens, retreat to the above law: discuss it with your supervisor. Resolving conflicts is part of every manager's job, your supervisor's included.

As a rule, you can serve all other ends to best advantage by assuming that your supervisor is approximately the right person for that job. It is not uncommon for young engineers, in their impatient zeal to get things done, to ignore, or attempt to go over or around their superiors. Sometimes they move a little faster that way, for a while, but sooner or later they find that such tactics cannot be tolerated in a large organization. Generally speaking, you cannot get by whoever evaluates your performance, for he or she rates you on your ability to cooperate, among other things. Besides, most of us get more satisfaction out of our jobs when we're able to display at least some personal loyalty to our superiors, with the feeling that we're helping them to get the main job done.

Be as particular as you can in the selection of your supervisor.

For most neophyte engineers, the influence of the senior engineers with whom they work, and even more so, the engineer to whom they report, is a major factor in molding their professional character. Long before the days of universities and textbooks, master craftsmen absorbed their skills by apprenticing to master craftsmen. Likewise, you will do well to use those with more experience, especially a well-selected supervisor, as your master, your mentor. A properly selected mentor will likely have been through gauntlets as severe as your present one, and will guide you through it much easier than you alone can.

But, of course, it is not always possible to choose a boss advisedly. What if yours turns out to be no more than half the supervisor you hoped for? There are only two proper alternatives open to you: (1) accept your boss as the representative of a higher authority and execute his or her policies and directives as effectively as possible, or (2) move to some other department, division, or company at the first opportunity. A great deal of mischief can be done to the interests of all concerned, including your company, if some other alternative is elected. Consider the damage to the efficiency of a military unit when the privates, disliking the leader, ignore or modify orders to suit their individual notions! To be sure, a business organization is not an army, but neither is it a mob.

Whatever your supervisor wants done takes top priority.

You may think you have more important things to do first, but unless you obtain permission it is usually unwise to put any other project ahead of a specific assignment from your own supervisor. As a rule, your boss has good reasons for wanting a job done now, and it is apt to have a great deal more bearing upon your performance rating than less conspicuous projects that may appear more urgent.

Whenever you are asked by your manager to do something, you are expected to do exactly that.

Whenever your supervisor sends you off to perform a specific task, you have two possible responses: (1) you do it exactly as requested, or (2) you come back and talk it over some more. (Take special note of this law, for it applies not only as regards your supervisor, but also to anyone with whom you have agreed on a task to be done or a course of action to be taken.) It is simply unacceptable either not to do it, or to do something different instead. If you become concerned that the planned action isn't worth doing as originally assigned (in view of new data or events), you may discuss, indeed you are obligated to discuss, the entire matter again, stating your intentions and reasons so that your manager can properly reconsider it.

Despite the responsibility to do exactly as instructed or agreed, you will sometimes want to prove your initiative by doing not only that, but also something in addition thereto; perhaps the next logical action has become clear, or perhaps a promising alternative has come to light. These can, within reason, be done in addition to the original assignment, and your drive and inventiveness will be immediately apparent.

Any violation of this law puts your trustworthiness at risk. Nevertheless, as with many of these laws, you will be forced to break this one on occasion, too. Do this only when you are certain that circumstances demand it (expediency being one such circumstance), and that the others involved will agree with your decision.

Do not be too anxious to defer to or embrace your manager's instructions.

This is the other side of the matter covered by the preceding two rules. An undue subservience or deference to any manager's wishes is fairly common among young engineers. Employees with this kind of philosophy may:

- plague their managers incessantly for minute directions and approvals,
- surrender all initiative and depend on their supervisor to do all the thinking for a project,
- persist with a design or a project even after new evidence has proven the original plan to be infeasible.

In general, a program laid down by the department, the project leader, or the design team is a proposal, rather than an edict. It is usually intended to serve only as a guideline, one that will have been formulated without benefit of the new information that will be discovered during its execution. The rule therefore is to keep others, your manager included, informed of what you have done, at reasonable intervals, and ask for approval of any well-considered and properly planned deviations that you may have conceived.

REGARDING RELATIONS WITH COLLEAGUES AND OUTSIDERS

Never invade the domain of any other department without the knowledge and consent of the manager in charge.

This is a common offense, which causes no end of trouble. Exceptions will occur in respect to minor details, but the rule applies particularly to:

- *The employment of a subordinate.* Never offer anyone a job, or broach the matter at all, without first securing the permission of his or her manager. There may be excellent reasons why that person should not be disturbed.

- *Engaging the time or committing the services of someone from a different department or division for some particular project or trip.* How would you feel, after promising in a formal meeting to assign one of your employees to an urgent project, to discover that someone else, without direct authority, has committed one of your subordinates to a task without attempting to notify you?

- *Dealings with customers or outsiders, with particular reference to making promises or commitments involving another department.* In this connection bear in mind that the manager of the department for which you would make such promises may have very good reasons not to want them made or not to be able to keep them. You simply do not have authority to commit other departments without the responsible manager's prior approval.

- *Performing any function assigned to another department or individual.* Violations of this law often cause bitter resentment and untold mischief. The law itself is based upon three underlying principles:

1. Most people strongly dislike having anyone "muscle" into their territory, undermining their job by appropriating their functions.

2. Such interference fosters confusion and mistakes. The individual who is in charge of the job usually knows much more about it than you do, and, even when you think you know enough about it, the chances are better than even that you'll overlook some important factor.

3. Whenever you are performing the other person's function you are probably neglecting your own. It is rare that engineers or executives are caught up enough with their own responsibilities that they can afford to take on those of their colleagues.

There is a significant commentary on this last principle that should also be observed: In general you will get no credit or thanks for doing the other person's job at the expense of your own. But it frequently happens that, if you can put your own house in order first, an understanding of and an active interest in the affairs of others will lead to promotion to a position of greater responsibility. More than a few employees have been moved up primarily because of a demonstrated capacity for helping to take care of other people's business as well as their own.

In all transactions be careful to "deal in" everyone who has a right to be in.

It is all too easy, especially in a large corporation, to overlook the interests of a department or individual who does not happen to be represented, or in mind, when a significant step is taken. Very often the result is that the step has to be retraced or else considerable damage is done. Even when it does no apparent harm, most people do not like to be left out when they have a stake in the matter, and the effect upon morale may be serious. Of course there will be crisis times when you cannot wait to stand on ceremony and you will have to steam full speed ahead with little regard for personal consequences. But you cannot do it with impunity too often.

Note particularly that in this and the preceding rule the chief offense lies in the invasion of someone else's territory without that person's knowledge and consent. You may find it expedient on occasion to do parts of other people's jobs in order to get your own work done, but you should first give them a fair chance to deliver on their own or else agree to have you take over. If you must offend in this respect, at least you should realize that you are being offensive.

Cultivate the habit of seeking other peoples' opinions and recommendations.

Particularly as a beginning engineer, you cannot hope to know all you must about your field and your employer's business. Therefore, you must ask for help from others; routinely seek out those who are "in the know."

This is particularly useful advice during a confrontation of any sort. A good first question to ask is: "What do you recommend?" Your confronter will usually have thought about it more than you have, and this will allow you to proceed to a productive discussion, and to avoid a fight.

A warning about soliciting others' opinions deserves mention. Condescending attitudes toward others and their opinions are gratuitous and unwelcome. If you have no intention of listening to, properly considering, and perhaps using someone's information or opinion, don't ask for it. Your colleagues will not take long to recognize such patronizing and to disdain you for it.

Promises, schedules, and estimates are necessary and important instruments in a well-ordered business.

Many engineers fail to realize this, or habitually try to dodge the responsibility for making commitments. You must make promises based upon your own estimates for the part of the job for which you are responsible, together with estimates obtained from contributing departments for their parts. No one should be allowed to avoid the issue by the old formula, "I can't give a promise because it depends upon so many uncertain factors." Consider the "uncertain factors" confronting a department head who must make up a budget for an entire engineering department for a year in advance! Even the most uncertain case can be narrowed down by first asking, "Will it be done in a matter of a few hours or a few months, a few days or a few weeks?" It usually turns out that it cannot be done in less than three weeks and surely will not require more than five, in which case you'd better say four weeks. This allows one week for contingencies and sets you a reasonable miss under the comfortable figure of five weeks. Both extremes are bad; good engineers will set schedules that they can meet by energetic effort at a pace commensurate with the significance of the job.

As a corollary of the foregoing, you have a right to insist upon having estimates from responsible representatives of other departments. But in accepting promises, or statements of facts, it is frequently important to make sure that you are dealing with a properly qualified representative. Also bear in mind that when you ignore or discount other engineers' promises you dismiss their responsibility and incur the extra liability yourself. Of course this is sometimes necessary, but be sure that you do it advisedly. Ideally, other engineers' promises should be reliable instruments in compiling estimates.

When you are dissatisfied with the service of another department, make your complaint to the individual most directly responsible for the function involved.

Complaints made to an individual's supervisors, over his or her head, engender strong resentment and should be resorted to only when direct appeal fails. In many cases such complaints are made without giving the individual a fair chance to correct the grievance, or even before he or she is aware of any dissatisfaction.

This applies particularly to individuals with whom you are accustomed to dealing directly or at close range, or in cases where you know the person to whom the function has been assigned. It is more formal and in some instances possibly more correct to file a complaint with the department head, and it will no doubt tend to secure prompt results. But there are more than a few individuals who would never forgive you for complaining to their supervisor without giving them a fair chance to take care of the matter.

Next to a direct complaint to the top, it is sometimes almost as serious an offense to send to a person's supervisor a copy of a document containing a complaint or an implied criticism. Of course the occasion may justify such criticism; just be sure you know what you are doing.

In dealing with customers and outsiders, remember that you represent the company, ostensibly with full responsibility and authority.

You may be only a few months out of college but most outsiders will regard you as a legal, financial, and technical agent of your company in all transactions, so be careful of your commitments.

PART II

RELATING CHIEFLY TO ENGINEERING MANAGERS

The following is a partial list of basic commandments, readily subscribed to by all managers, but practiced only by the really good ones.

INDIVIDUAL BEHAVIOR AND TECHNIQUE

Every manager must know what goes on in his or her domain.

This is repeated here for emphasis, and because it belongs at the head of the list for this section. Just remember that it works both ways, as regards what you owe your associates and subordinates as well as yourself.

Obviously this applies primarily to major or significant developments and does not mean that you should attempt to keep up with all the minor details of functions assigned to subordinates. It becomes a vice when carried to the extent of impeding operations. Nevertheless, the basic fact remains that the more information managers have, the more effectively they can manage their business.

Do not try to do it all yourself.

This is another one of those elementary propositions that everyone will endorse and yet many will carelessly violate. It's bad business: bad for you, bad for the job, and bad for your employees. You must delegate responsibility even if you could cover all of the ground yourself. It isn't wise to have so much depend upon one person, and it's unfair to your subordinates. Executives should have their business organized so that they could be away on business or vacation at any time and still have everything go along smoothly. The most common excuse for hogging the whole job is that subordinates are too young or inexperienced. It's part of your job to develop your subordinates, which includes developing initiative, resourcefulness, and judgment. The best way to do this is to load them up with all the responsibility they can carry without danger of serious embarrassment to any person or group. Self-respecting engineers resent being babied to an extent where they cannot act on even the most trivial detail without express approval of their manager.

On the other hand, it must be granted that details are not always trivial, and that it may sometimes require a meeting of an executive committee to change the length of a screw in a critically important mass-produced assembly. It is a matter of making sure not only that technical items are handled by engineers of appropriate competence and experience but that all considerations have been made (Rabbe).

Put first things first in applying yourself to your job.

Since there usually isn't time for everything, it is essential to form the habit of concentrating on the important things first. The important things are the things for which you are held directly responsible and accountable, and if you aren't sure what these are you'd better find out mighty quick and fix them clearly in your mind. Assign these responsibilities top priority in budgeting your time; then delegate as many as possible of the items that will not fit into your schedule. It is a good general rule never to undertake any minor project or chore that you can get someone else or some other department to do for you, so long as it is not an essential part of your job. For example, if your job is building motors it is a mistake to spend time designing special vibration or sound meters for testing them if you can get a laboratory or service to do it for you.

The practice of drawing upon all available resources for assistance can frequently be applied to advantage in respect to your major products, as well as in minor details. This is especially true in a large organization where the services of experts, consulting engineers, laboratories, and other departments are available; they will almost always be able to get an answer far more efficiently than you could independently. In fact, there will be cases in which it would be wise for you to limit yourself, personally or as a business manager, to performing only those functions to which you can bring some special talent, skill, or contribution, or in which you enjoy some natural advantage. The common belief that everyone can do anything if they just try hard enough is a formula for inefficiency at best and for complete failure at worst. Few of us are versatile enough to excel in more than a very few talents.

Cultivate the habit of "boiling matters down" to their simplest terms.

The faculty for reducing apparently complicated situations to their basic, essential elements is a form of wisdom that must usually be derived from experience, but there are marked differences between otherwise comparable individuals in this respect. Some people seem eternally disposed to "muddy the water"; or they "can never see the forest for the trees." Perhaps one cannot correct such an innate tendency simply by taking thought, but it appears to be largely a habit — a habit of withdrawing mentally to a suitable vantage point so as to survey a mass of facts in their proper perspective, or a habit of becoming immersed and lost in a sea of detail. Make it a practice to integrate, condense, summarize, and simplify your facts rather than expand, ramify, complicate, and disintegrate them.

Many meetings, for example, get nowhere after protracted wrangling until somebody finally says "Well, it all boils down simply to this . . .," or "Can't we agree, however, that the basic point at issue is just this . . .,", or "After all, the essential fact remains that . . . ." This sort of mental discipline, which instinctively impels one to get at the heart of the matter, is one of the most valuable qualities of a good executive (Smith).

Do not get excited in engineering emergencies — keep your feet on the ground.

This is certainly clear enough, and yet organizations will sometimes be thrown into a state of agitation bordering on panic by some minor crisis. This refers especially to bad news from the factory or the field regarding some serious and embarrassing difficulty, such as an epidemic of equipment or product failures. Most crises aren't half as bad as they appear at first, so make it a point not to magnify a bad situation. Do not ignore signs of trouble and get caught napping, but, learn to distinguish between isolated cases and real epidemics. To be sure, hazards — human safety or environmental — rate an immediate, aggressive response; potential liability demands it. Nevertheless, in any case, the important thing is to get the facts first, as promptly and as directly as possible. Then act as soon as you have enough evidence from responsible sources to enable you to reach a sound decision.

Engineering meetings should neither be too large nor too small.

Some managers carry their aversion for large meetings to the point of phobia. It is true enough that large meetings frequently dissipate the subject over a number of conflicting or irrelevant points of view in a generally superficial manner. But this is almost entirely because of the competence of the person leading the meeting. A considerable amount of skill is required to manage a sizable meeting so as to keep it on the proper subject, avoiding long-winded digressions or reiterations of the arguments. It should be the function of the chairperson, or the presiding manager, to bring out the pertinent facts bearing upon the matter, in their logical order, and then to secure agreement upon the various issues by asking for general assent to concrete proposals, taking a vote, or making discretionary decisions. Engineering meetings may degenerate into protracted quarrels for lack of competent direction. This danger seems to be proportionate to the size of the meeting.

Small meetings (three or four persons) can usually hammer out a program or dispose of perplexing problems much more effectively. The chief drawback lies in the very serious possibility that all interested parties may not be represented, and considerable mischief may result from failing to take account of significant facts or points of view. Apart from any actual loss involved, strong resentment or discouragement may be aroused in the neglected parties.

There will doubtless be cases in which it is neither feasible nor desirable to have all interested parties represented in engineering discussions, particularly if the participants are well informed. But in general it is fitting, proper, and helpful to have present those whose particular territory is under discussion. An excellent means for avoiding the objections to either extreme in this respect is to keep the meeting small, calling in key individuals when their particular responsibilities are being discussed.

Along this same line, a worthwhile guideline to observe is to limit attendees to two levels of the organizational structure, especially for working, decision-making meetings (informational meetings are another matter). A third or fourth level in attendance has no decision-making power anyway, practically speaking. Aside from the exception of presenting information their presence is usually wasteful. On the other hand, if lower levels can make decisions on the items being discussed, attendance by the higher level people is even more wasteful; higher-up participants often detrimentally dominate the discussion and decisions.

In any meeting the important thing is to face the issues and dispose of them. All too often there is a tendency to dodge the issues, postponing action until a later date, or "letting the matter work itself out naturally." Matters will always work out "naturally" if the executive function of control is neglected, but this represents a low order of management. Count any meeting a failure that does not end up with a definite understanding as to what's going to be done, who's going to do it, and when, and this should be confirmed in written minutes.

Cultivate the habit of making brisk, clean-cut decisions.

This is, of course, the most difficult and important part of a manager's job. Some managers have a terrific struggle deciding even minor issues, mainly because they never get over being afraid of making mistakes. Normally, facility comes with practice, but it can be hastened by observing a few simple principles.

1. Decisions will be easier and more frequently correct if you have the essential facts at hand. It will therefore pay you to keep well informed, or else to bring out the relevant facts before attempting a decision. However, it is sometimes said that anybody can make decisions when all of the facts are at hand,

whereas a good manager will make the same decisions without waiting for the facts (Osborne). To maintain a proper balance in this respect, when in doubt ask yourself: "Am I likely to lose more by giving a snap judgment or by waiting for more information?"

2. The application of judgment can be facilitated by formulating it into principles, policies, and precepts in advance. The present chapter is an attempt to formulate experience for this purpose. Make up your own code, if you will, but at least have some sort of code, for much the same reason that you memorize physical constants or Newton's laws of motion.

3. You do not have to be right every time. It is said that a good executive needs to be right only 51 percent of the time (although a better margin will serve you proportionately better).

4. The very fact that a decision is difficult usually means that the advantages and drawbacks of the various alternatives are pretty well balanced, so that the net loss cannot amount to much in any event. In such cases it is frequently more important to arrive at some decision — any decision — promptly than to arrive at the best decision ultimately. So take a definite position and see it through.

5. It is futile to try to keep everybody happy in deciding issues involving several incompatible points of view. By all means give everyone a fair hearing, but after all parties have had their say and all facts are on the table, dispose of the matter decisively even if someone's toes are stepped on. Otherwise the odds are that everyone will end up dissatisfied, and even the chief beneficiary will think less of you for straddling the issues.

The following criteria are helpful in choosing a course of action when the factors are indecisive; ask yourself these questions:

• Does it expedite and forward the undertaking, or does it only produce procrastination and delay?

• Is it fair and square and aboveboard?

• Is it in line with established custom, precedence, or policy? A good reason is generally required for a departure.

- Is it in line with a previous specific decision or understanding? Even a good reason for making a change will sometimes not offset the unfortunate impression of apparent instability. "He can't make up his own mind" is a common reaction. (Observe, however, that this criterion is suggested only "when other factors are indecisive." By all means have the courage of your convictions when the change is justifiable.)
- What are the odds? Can we accept the risk? How does the possible penalty compare with the possible gain in each of the alternatives offered? Very often you can find a solution wherein the worst possible consequence isn't too bad, in relation to the possible gains.

Do not allow the danger of making a mistake to inhibit your initiative to the point of "nothing ventured, nothing gained." It is much healthier to expect to make mistakes, take a few good risks now and then, and take your medicine when you lose. Moreover, there are few mistakes that cannot be turned into profit somehow, even if only for the experience.

On the opposite side, never mistake the true meaning of the statement, "Don't be afraid to make mistakes." Incorrect decisions that result in catastrophic consequences such as huge financial losses or personal injury will not be overlooked. Such decisions may even be criminal conduct. Those who coerce you into making decisions, despite whatever aphorisms they may employ toward that end, do not expect and will not accept a catastrophic result. Indeed, don't be afraid to make mistakes; make clear-cut, swift decisions, but only if a mistake won't create wreckage for your organization — and you.

Do not overlook the value of suitable "preparation" before announcing a major decision or policy.

When time permits, it is frequently good diplomacy to prepare the ground for such announcements by discussing the matter in advance with various key personnel or directly interested parties.

This is, in fact, an elementary technique in diplomatic and political procedure, but it is all too often ignored in engineering practice. Much embarrassment and bad feeling can be caused by announcing a major change or embarking upon a new program or policy without consulting those directly affected or those who are apt to bring up violent objections later on (Schell).

MANAGING DESIGN AND DEVELOPMENT PROJECTS

Learn project management skills and techniques, then apply them to the activities that you manage.

Your organization probably has, or certainly should have, standard procedures for its major engineering efforts such as developing new products or processes. You will also need to apply techniques commonly used for managing projects. Some of these are, for example, resource planning, calendar scheduling, and progress tracking. Simply stated, to manage projects properly you must plan your work, then work your plan.

The following formula for carrying out any engineering project seems to be more or less standard in the best engineering circles (Project Management Institute):

1. Define your project's objectives.
2. Plan the job by:
   - outlining the steps to be accomplished;
   - defining the required resources, including people, money, and facilities;
   - preparing a definite schedule.
3. Execute the plan.
4. Monitor the progress and respond to deviations.
   - Watch for "bottlenecks," "log-jams," and "missing links"; hit lagging items hard by applying additional time, money, and people.

- Revise your schedule as required.
5. Drive to a finish on time.

Plan your development work far enough ahead of production so as to meet schedules without a wild last-minute rush.

Although this rule apples to any activity just as well as it does to production planning, this seems to be a particularly common — and devilish — error. Although not a hard and fast rule, in many organizations, it is common that someone responsible for design engineering is also the one to manage the overall development project. Everyone, designers included, has the natural tendency to become preoccupied with one's own problems and areas of expertise, and to underestimate or even ignore those of other departments or disciplines. Considerable foresight is required to offset this natural tendency, for inadequate planning and execution of the "whole" project is poor project management. Even when a new design is no more than old fundamentals in a new cloak, it is important to plan the program early enough and to provide for all stages in the process of getting the product on the market. Items or activities as remote from design engineering as inventory planning, regulatory approval, user's instructions, or promotional materials are every bit as important to successfully completing product development as are design engineering, process development, or assembly fixtures.

Beware of seeking too much comfort in planning your engineering programs.

Too much preoccupation with the pursuit of security is apt to lead to greater danger and insecurity. In a competitive world you must take chances — bold and courageous chances — or else others will, and they will win out just often enough to keep you running, all out of breath, trying to catch up. So it behooves you as an engineering manager to undertake stiff development programs, thereby setting a high mark, and then working aggressively to meet it. With competent direction any representative engineering organization will work its way out of a tight spot under the pressure of the emergency. If you do not create your own emergencies in advance, your competition will create them for you at a much more embarrassing time later on.

In order to minimize the risk inherent in aggressive programs (technical or schedule challenges included), it is good policy to hedge against failure by providing an alternative, or an "out" to fall back on, wherever practicable. You can go after bigger stakes with impunity when you have suitably limited your possible losses in such a manner.

Be content to "freeze" a new design when the development has progressed far enough.

Of course it is not always easy to say how far is "far enough," but in general you have gone far enough when you meet the design specifications and budget, with just enough time left to complete the remainder of the program on schedule. The temptation of design engineers is to allow themselves to be led on by one glittering improvement after another, pursuing an elusive perfection that takes them far past the hope of ever keeping their promises and commitments. Bear in mind that there will always be new design improvements coming along, but it is usually better to get started with what you have developed on time, provided only that it is up to the specifications for features, quality, and cost.

Constantly review projects to make certain that actual benefits are in line with costs in money, time, and human resources.

Not infrequently projects are carried along by virtue of Newton's first law of motion long after they could ever yield a satisfactory return on the investment. The occasion for vigilance in this respect is obvious enough; it is cited here simply as a reminder.

Make it a rule to require, and submit, regular periodic progress reports, as well as final reports on completed projects.

However tedious such chores may seem, your business simply isn't fully organized and controlled until you have established this practice, as regards reports to your superiors as well as from your subordinates. There appears to be no other instrument quite so compelling and effective in requiring an engineer to keep the facts properly assembled and appraised. Such reports are, therefore, equally as useful for the writers as the readers.

It is further true that, generally speaking, an engineering project is not really finished until it is properly summarized, recorded, and filed in such a manner that the information can readily be located and used by all interested parties. Most organizations have clear policies on part drawings and production specifications, but good development organizations would never overlook documenting engineering and laboratory knowledge as well. An enormous amount of effort can be wasted or duplicated in an engineering department when this sort of information is simply entrusted to the memory of individual engineers.

ON ORGANIZATIONAL STRUCTURES

Make sure that everyone has been assigned definite positions and responsibilities within the organization.

It is extremely detrimental to morale and efficiency when employees do not know just what their jobs are or what they are responsible for. If assignments are not made clear there is apt to be interminable bickering, confusion, and bad feeling. Do not keep tentative organization changes hanging over people; effect them as soon as they become reasonably clear. Changing them again later is better than leaving people in poorly or improperly defined positions.

All employees, engineers included, can be organizationally linked to one another based on their project (e.g., new product development team, program implementation team), their functional discipline (e.g., stress analysis, analog circuit design, R&D), or both. The last, also called a matrix organization, gives to each person (at least) two managers: one for a project and one for a discipline. The functional discipline supervisor usually has administrative authority — performance appraisals, promotions, compensation — over the employee. This type of organization affords the great advantage to everyone of having two supervisors from which to obtain help; but it suggests the possible disadvantage of competition for allegiance. Well-managed organizations will not suffer; conflicts will be easily resolved by considering the grander goals, preferably, or by the next level of management.

Make sure that everyone has the authority they need to execute their jobs and meet their responsibilities.

This is commonly expressed in the saying that authority must be commensurate with responsibility. Ideally each individual should have full authority and control over all of the factors (budgets, expenditures, and personnel) essential to the performance of his or her particular job. In practice this is seldom achievable; we must all depend upon the willing contributions of others at some point in the process. Still, the amount of dependency should be kept to the practical minimum, for it is extremely difficult for employees to get anything done if they must eternally solicit the voluntary cooperation or approval of too many other parties.

Bestowing or acquiring the needed authority is difficult and delicate. A fine line exists between being authoritative and being overbearing, but a clear line exists between the resulting effectiveness or impotence. The important thing is to exercise sufficient care to avoid running afoul of the interests and authority of others.

In reality, employees will frequently be held responsible for a good deal more than they can control by directly delegated authority. Nonetheless, a considerable amount of authority can be assumed with complete impunity if it is assumed discreetly, and with effective results. In general we tend to obey people who appear to be in charge of any situation, provided that they appear to know what they are doing and obtain the desired results. Most managers, yourself included, should be pleased to confirm such authority in their subordinates when they see it being exercised effectively. Therefore, you should encourage yourself and your subordinates to assume, albeit delicately, however much authority is needed to do the job.

Make sure that all activities and all individuals are supervised by someone competent in the subject matter involved.

At high levels in a chain-of-command, supervisors and subordinates will oftentimes have dissimilar training and experience. But in an engineering organization, at least ideally, every novice engineer working in a technological area should be supervised by a veteran seasoned in the same area. Neophytes can get themselves, their departments, their employers, and their supervisors into embarrassingly difficult spots if left on their own.

As a manager, you should offer your subordinates complete competence and instill in them the same when overseeing them in their technical activities. If you are uncomfortable with this responsibility, you'd better do something about it, and with all due haste. A particularly good method to use, outside of directly learning what you need to know, is to complement yourself with other experienced people under your supervision, people who can properly judge what you cannot, thereby shoring up any breaks in your expertise. Another method is to assign technical as well as project responsibilities to every engineer in the organization, and then have the technical experts act as consultants for others. Matrix organizations can excel at assuring such interactions.

As a final consideration, recognize the limitations regarding how many subordinates one can adequately manage. The conventional guideline, in keeping with traditional organizational theory, is not to have more than six or seven people report to one supervisor. Adhering to a rigid standard in this regard is excessive, what with organizations and individuals each being unique, but never deprive an employee of adequate supervision because of a lopsided reporting structure.

WHAT ALL MANAGERS OWE THEIR EMPLOYEES

Never misrepresent a subordinate's performance during performance appraisals.

The most serious responsibility of managers is to review the performance of their subordinates. As a manager, you have the distinct obligation to do this as accurately as possible. Not only would misrepresentation be unfair to your subordinates, it would not be the least bit helpful to anyone involved.

Along the same line, it is your inalienable responsibility to talk things over with employees if — and as soon as — you become sincerely dissatisfied with their work, or you recognize deficiencies that are working against them. To be sure, this is not always easy, and it will require much tact to avoid discouraging or offending them, but you owe it to them. Bear this in mind: If you ultimately must fire a subordinate, you may have two pointed questions to answer: "Why has it taken you five years to discover my incompetence?" and "Why haven't you given me a fair chance to correct these shortcomings?" Remember that when you fire someone for incompetence, it means not only that the employee has failed, but also that you have failed.

Make it unquestionably clear what is expected of employees.

Number one on the list of required communication between supervisor and subordinate is the explicit understanding of expectations on the job. All too often, managers avoid direct discussions and rely on implicit instructions, generalized goals, or corporate policies. It is not enough for you simply to hope for certain behavior or performance from your subordinates; more often than not you will be disappointed. Successful managers clearly set down goals and expectations with their subordinates, then follow up with monitoring and support.

Promote the personal and professional interests of your employees on all occasions.

This is not only an obligation — it is the opportunity and the privilege of every manager.

The interests of individual engineers coincide at least in principle with the company's interest, meaning that there is, or should be, no basic conflict. The question of which should be placed first is, therefore, rarely encountered in practice, although it is clear that, in general, the company's interests, like those of the state or society, must take precedence. It is one of the functions of management to reconcile and merge the two sets of interests to their mutual advantage, since they are so obviously interdependent.

Clearly, it is to the company's advantage to preserve the morale and loyalty of individual engineers. These are tremendously important factors in any organization. They are founded primarily upon confidence, and reach a healthy development when employees feel that they will always get a square deal, plus a little extra consideration on occasions.

Do not hang on to employees too selfishly when they are offered a better opportunity elsewhere.

It's bad business to stand in the way of a subordinate's promotion just because the loss will inconvenience you. You are justified in shielding your people from outside offers only when you are sincerely convinced that they have an equal or better opportunity where they are. Accept that you are probably unable to judge this yourself anyway, so consider soliciting the opinion of the employee involved; it is his or her career, not yours. Anyway, you should not get caught in a position where the loss of an individual will embarrass you unduly. Select and train back-ups for all key personnel, including yourself.

Do not short-circuit or override your subordinates if you can possibly avoid it.

It is natural, on occasion, for a manager to want to exercise managerial authority directly in order to dispose of a matter promptly without regard to the engineer assigned to the job. To be sure, it's your prerogative, but it can be very demoralizing to the subordinate involved and should be resorted to only in real emergencies. Once you assign jobs to your people, let them do those jobs, even at the cost of some inconvenience to yourself. Finally, you can do irreparable damage by exercising authority without sufficient knowledge of the details of the matter.

You owe it to your subordinates to keep them properly informed.

In the catalog of raw deals, next to responsibility without authority comes responsibility without information. It is very unfair to ask engineers to acquit themselves creditably when they are held responsible for a project without having adequate knowledge of its past history, present status, or future plans. An excellent practice is to hold occasional meetings to acquaint employees with major policies and developments in the business of the department and the company.

An important part of the job of developing engineers is to furnish them with ample background knowledge in their particular domain, and as a rule this involves a certain amount of travel. There are occasions when it is worthwhile to send young engineers along on trips for what they can get out of a job, regardless of how little they can directly contribute. Likewise, include interested individuals in introductions, luncheons, and so on, when hosting visitors. Obviously, this can be overdone, but when outsiders visit, it is good business, as well as good manners, to invite all the involved internal persons to participate as well.

Do not criticize a subordinate in front of others, especially his or her own subordinates.

This will damage both prestige and morale. Also, be very careful not to criticize someone when it is really your own fault. Not infrequently, the real offense can be traced back to you, as when you failed to advise, or warn, or train the individual properly. Be fair about it.

Show an interest in what your employees are doing.

It is discouraging to engineers when the boss manifests no interest in their work, as by failing to inquire, comment, or otherwise take notice of it. A little effort goes a long way — make the effort.

Never miss a chance to commend or reward subordinates for a job well done.

Remember that your job is not just to criticize your people and intimidate them into getting their work done. A first-rate manager is a leader as well as a critic. The better part of your job is, therefore, to help, advise, encourage, and stimulate your subordinates. Along the same line, never miss a chance to build up the prestige of your subordinates in the eyes of others.

On the other hand, this is not to suggest perpetual lenience. By all means get tough when the occasion justifies it. An occasional sharp censure, when it is well deserved, will usually help to keep employees on their toes. But if that's all they get, they are apt to go a bit sour on the job.

Always accept full responsibility for your group and the individuals in it.

Never "pass the buck," or blame any of your employees, even when they may have "let you down" badly. You are supposed to have full control and you are credited with the success as well as the failure of your group.

Do all you can to see that your subordinates get all of the salary to which they are entitled.

Undeniably, we work in large part because we are paid for it. Pay increases, however they are manifested, are the most appropriate reward or compensation for outstanding work, greater responsibility, or increased value to the company. (Any recommendation or an increase in compensation must be justified on one of these three bases.)

Do all you can to protect the personal interests of your subordinates and their families.

You needn't confine your interest in your employees rigidly within the boundaries of company business. Most will appreciate your honest, unprying interest in their lives outside of the workplace, and if they have personal difficulties, your support as well. Be sure, above all, to respect the family and religious desires and obligations of your employees.

Try to get in little extra accommodations when justifiable. For example, if you're sending an engineer to his or her home town on a business trip, allow some schedule flexibility so that the employee can spend some free time there, if it makes no difference otherwise. Considerations of this sort make a nice difference in the matter of morale and in the satisfaction a manager gets out of the job. Treat your people as human beings making up a team rather than as cogs in a machine.

To be sure, managers have the right to inconvenience their subordinates, but they should also embrace the responsibility to avoid doing so whenever possible. Managers who insist, at every opportunity, upon flagrantly displaying power over their subordinates by disrupting their personal lives can expect bitterness and resentment in return.

PART III

PROFESSIONAL
AND PERSONAL
CONSIDERATIONS

A number of empirical studies of on-the-job excellence have clearly and repeatedly established that emotional competencies — communication, interpersonal skills, self-control, and so on — "play a far larger role in superior job performance than do cognitive abilities and technical expertise" (Goleman, p. 320). Yet most of the emphasis in the education and training of engineers is placed upon purely technical education.

Notwithstanding some brilliant exceptions, intelligence, academic training, technical knowledge, and circumstantial expertise alone are not major determinants in the success or failure of engineers in the workplace. For the most part, engineers are or can quickly become adequately capable in these areas. If technically incapable, they almost certainly would have been discharged from the system, either by themselves or by others, long before they became employed engineers. Generally, such skills and traits as communication, confidence, group and interpersonal effectiveness, motivation, pride in accomplishments, adaptability, leadership potential, inquisitiveness, integrity, and emotional control are exhibited by the most successful employees, just as with the most successful among engineers.

It should be obvious enough that a highly trained technical expert with a good character and personality is necessarily a better engineer and a great deal more valuable as an employee than a sociological freak or misfit with the same technical training. This is largely a consequence of the elementary observation that in a normal organization one cannot get very far in accomplishing anything worthwhile without the voluntary cooperation of one's associates; and the quantity and quality of such cooperation is determined by the "personality factor" as much as anything else. Added to this need for one-on-one cooperation are all sorts of "soft" characteristics from understanding contemporary society to following ethical behavior — all of which can add up to benefits for yourself and your employer far beyond ordinary technical contributions.

The following "laws" are drawn up from the purely practical point of view. As in the two preceding sections, the selections are limited to rules that are frequently violated, with unfortunate results, however obvious or stale they may appear.

LAWS OF CHARACTER AND PERSONALITY

One of the most valuable personal traits is the ability to get along with all kinds of people.

This is rather a comprehensive quality but it defines the prime requisite of personality in any type of industrial organization. No doubt this ability can be achieved by various formulas, although it is probably based mostly upon general, good-natured friendliness, together with fairly consistent observance of the "Golden Rule." The following "do's and don'ts" are more specific elements of such a formula.

1. *Cultivate the tendency to appreciate the good qualities, rather than the shortcomings, of each individual.*

2. *Do not give vent to impatience and annoyance on slight provocation.* Some offensive individuals seem to develop a striking capacity for becoming annoyed, which they indulge with little or no restraint.

3. *Do not harbor grudges after disagreements involving honest differences of opinion.* Keep your arguments objective and leave personalities out as much as possible. You never want to create enemies, for as E. B. White put it: "One of the most time-consuming things is to have an enemy."

4. *Form the habit of considering the feelings and interests of others.*

5. *Do not become unduly preoccupied with your own selfish interests.* It is natural enough to "look out for Number One first," but when you do, your associates will be noticeably disinclined to look out for you, because they know you already are. This applies particularly to the matter of credit for accomplishments. It is much wiser to give your principal

attention to the matter of getting the job done, or to building up your associates than to spend too much time pushing your personal interests ahead of everything else. You need have no fear of being overlooked; about the only way to lose credit for a creditable job is to grab for it too avidly.

6. *Make it a rule to help the other person whenever an opportunity arises.* Even if you're mean-spirited enough to derive no personal satisfaction from accommodating others, it's a good investment. The business world demands and expects cooperation and teamwork among the members of an organization. It is smarter and more pleasant to give it freely and ungrudgingly (up to the point of unduly neglecting your own responsibilities).

7. *Be particularly careful to be fair on all occasions.* This means a good deal more than just fair upon demand. All of us are frequently unfair, unintentionally, simply because we do not habitually view matters from other points of view to ensure that the interests of others are fairly protected. For example, an individual who fails to carry out an assignment is sometimes unjustly criticized when the real fault lies with the manager who failed to provide the tools to do the job. Whenever you enjoy some natural advantage, or whenever you are in a position to mistreat someone seriously, it is especially incumbent upon you to "lean over backwards" to be fair and square.

8. *Do not take yourself or your work too seriously.* A normal healthy sense of humor, under reasonable control, is much more becoming, even to an executive, than a chronically sour dead-pan, a perpetually unrelieved air of tedious seriousness, or a pompous righteousness. It is much better for your blood pressure, and for the morale of the office, to laugh off an awkward situation now and then than to maintain a tense, tragic atmosphere of stark disaster whenever matters take an embarrassing turn. To be sure, a serious

matter should be taken seriously, and employees should
maintain a quiet dignity as a rule, but it does more harm
than good to preserve an oppressively heavy and funereal
atmosphere around you.

9. *Put yourself out just a little to be genuinely cordial in greet-
ing people.* True cordiality is, of course, spontaneous and
should never be affected, but neither should it be inhibited.
We all know people who invariably pass us in the hall or
encounter us elsewhere without a shadow of recognition.
Whether this is due to inhibition or preoccupation, we cannot help feeling that such unsociable chumps would not be
missed much if we just didn't see them. On the other hand,
it is difficult to think of anyone who is too cordial, although
it can doubtless be overdone like anything else. It appears
that most people tend naturally to be sufficiently reserved or
else, as is common with engineers, over-reserved.

10. *Give people the benefit of the doubt if you are inclined to
suspect their motives, especially when you can afford to do
so.* Mutual distrust and suspicion breed a great deal of
absolutely unnecessary friction and trouble, frequently of a
very serious nature. It is derived chiefly from misunder-
standings, pure ignorance, or from an ungenerous tendency
to assume that people are guilty until proven innocent. No
doubt the latter assumption is the "safer" bet, but it is also
true that if you treat others as depraved scoundrels, they will
usually treat you likewise, and they will probably try to live
down to what is expected of them. On the other hand you
will get much better cooperation from your associates and
others if you assume that they are just as intelligent, reasonable, and decent as you are, even when you know they are
not (although setting the odds of that are tricky indeed!). Do
not fear being taken as naive or gullible; you'll gain more
than you lose by this practice with anything more than casu-
al attention to the actual odds in each case.

Do not be too affable.

It is a mistake, of course, to try too hard to get along with everybody merely by being agreeable or even submissive on all occasions. Somebody will take advantage of you sooner or later, and you cannot avoid trouble simply by running away from it. Do not give ground too quickly just to avoid a fight, when you know you're in the right. If you can be pushed around easily the chances are that you will be pushed around. Indeed, you can earn the respect of your associates by demonstrating your readiness to engage in a good (albeit non-personal) fight when your objectives are worth fighting for. Shakespeare put it succinctly in Polonius's advice to his son (in *Hamlet*): "Beware of entrance to a quarrel, but being in, bear 't that the opposed may beware of thee."

Like it or not, as long as you're in a competitive business you're in a fight; sometimes it's a fight between departments of the same company. As long as it's a good clean fight, with no hitting below the belt, it's perfectly healthy. But keep it to "friendly competition" as long as you can. In the case of arguments with your colleagues, it is usually better policy to settle your differences "out of court," rather than to take them higher for arbitration.

Likewise, in your relations with subordinates it is unwise to carry friendliness to the extent of impairing discipline. Every one of your employees should know that whenever they deserve a reprimand, they'll get it, every time. The most rigid discipline is not resented so long as it is reasonable, impartial, and fair, especially when it is balanced by appropriate praise, appreciation, and compensation. At the extreme, there may well be times when firing or transferring an employee is the best course of action, both for the person involved and for the company. If you do not face your issues squarely, someone else will be put in your place who will.

Regard your personal integrity as one of your most important assets.

In the long run there is hardly anything more important to you than your own self-respect, and this alone should provide ample incentive to maintain the highest standards for honesty and sincerity of which you are capable. But, apart from all considerations of ethics and morals, there are perfectly sound business reasons for conscientiously guarding the integrity of your character.

The integrity to which we refer is easily described: if you have high personal integrity, you are honest, morally sound, trustworthy, responsible, and sincere. The priceless and inevitable reward for uncompromising integrity is confidence: the confidence of associates, subordinates, and outsiders. All transactions are enormously simplified and facilitated when your word is as good as your bond and your motives are above question. Confidence is such an invaluable business asset that even a moderate amount of it will easily outweigh any temporary advantage that might be gained by having lost it.

A most striking phenomenon in an engineering office, once it is pointed out, is the transparency of character among members who have been associated for any length of time. In a surprisingly short period individuals are recognized, appraised, and catalogued for exactly what they are, with far greater accuracy than they usually realize. This makes anyone appear downright ludicrous when assuming a pose or trying to convince us that they are something other than they are. As Ralph Waldo Emerson puts it: "What you *are* . . . thunders so that I cannot hear what you say to the contrary." Therefore, it behooves you to let your personal conduct, overtly and covertly, represent the very best practical standard of personal and professional integrity by which you would like to let the world judge and rate you.

Moreover, it is morally healthy and creates a better atmosphere if you will credit the other person with similar standards of integrity. The obsessing and overpowering fear of being cheated is a common characteristic of substandard personalities. This sort of psychology sometimes leads one to credit oneself with being impressively clever when one is simply taking advantage of those who are more considerate and fair-minded.

In order to avoid any misunderstanding, it should be granted here that the average person, and certainly the average engineer, is by no means a low, dishonest scoundrel. In fact, the average person would violently protest any questioning of his or her essential honesty and decency, perhaps fairly enough. But the average person will often compromise whenever it becomes moderately uncomfortable to live up to his or her obligations. This is hardly what is meant by integrity, and it is difficult to base even moderate confidence upon the guarantee that you will not be deceived — unless the going gets tough.

Never underestimate the extent of your professional responsibility and personal liability.

Upon becoming a member of the engineering profession, you accepted the responsibility of a professional as well as any liability that accompanies that responsibility, be it personal, professional, or corporate. Many engineers pretend that they can hide behind their employer's or their department's shield, or that they are powerless, mere cogs in the machinery, even if, or especially if, something goes haywire. Although environmental, consumer, and product safety concerns are every individual's responsibility, an engineer, perhaps more than anyone, is uniquely positioned with the power and knowledge to create, identify, avoid, and correct such problems — an incongruous reality. Regardless of the size of your employer, never forget that, in the end, you contribute to making decisions, whether the results are good, bad, or catastrophic.

All this responsibility and liability will tempt you towards, but must not lead you to inaction and indecision. You needn't be unreasonably anxious; you are in your position as an engineer presumably because you have or can somehow bring to bear the training, knowledge, and experience to identify and judge the risks inherent in doing your business. Do your job responsibly, and you minimize liability all around. In this regard you will serve yourself, your employer, and your profession well if you follow a few simple guidelines:

- Approach all of your engineering systematically, especially when developing new products, processes, or equipment.

- Identify and apply the requisite expertise to all engineering activities.
- Be aware of and use applicable codes and standards.
- Use established procedures for design reviews and failure analyses.
- Keep records of your and your department's engineering activities.

Let ethical behavior govern your actions and those of your company.

Despite the usual ambiguities and everyday quandaries of engineering, ethical behavior usually comes naturally to engineers. Societal values — the basis for ethics — are ingrained. On the other hand, many times the ethical problems encountered in engineering practice are very complex and involve conflicting ethical principles (Fleddermann, p. 3). The accounts of many well-known cases of injurious product failures and tragic man-made disasters prove the point.

It should be observed that having the courage of your convictions includes having the courage to do what you know to be right, technically as well as ethically and morally, without undue fear of possible criticism or of the need to explain your actions. If your reasons for your actions are sound, you should not worry about having to defend them to anyone; if they're not sound you'd better correct them promptly, instead of building up an elaborate camouflage.

Understand, you are ill-advised to martyr yourself for every controversial matter in which you strongly believe. Martin Luther King, Jr. said: "If a man hasn't discovered something that he will die for, he isn't fit to live." True enough, but Oscar Wilde said: "A thing is not necessarily true because a man dies for it." Martyrdom only rarely makes heroes, and in the business world, such heroes and martyrs alike often find themselves unemployed.

Knowing what is ethically right, both for you and for your company, and then acting appropriately is the key. Professional societies offer good places to start with their codes of ethics, some of which are appended in this book.

REGARDING BEHAVIOR IN THE WORKPLACE

Be aware of the effect that your personal appearance has on others and, in turn, on you.

Permissiveness and dress codes aside, your appearance probably has a far greater influence on how you are viewed by those around you than you could ever imagine. Bear this in mind when you define and present your workplace image. Three rules of thumb will serve you well in this regard.

1. Look at how those in the positions to which you aspire are dressed and groomed, then follow their lead.

2. Dress appropriately for the occasion, whatever it is, including everyday work. When in doubt, slightly overdressing is prudent; being noticeably underdressed, at least for most people most of the time, is unbearably uncomfortable.

3. Conservative styles and colors in clothing as well as conservative grooming will never be wrong, at least in most engineering circles.

Despite the wide range of acceptable personal appearance within any one office, and the vastly greater range to be found in society overall, it's hard to argue against these common-sense points.

• Clothing, regardless of style, should be clean, well-fitting, and in good condition.

• Hair and nails should be clean and well-kept, again regardless of style.

• Your good personal hygiene will be appreciated by your colleagues.

• Perfumes and colognes should be used sparingly, if at all, in the workplace.

• Men should pay particular attention to shaving habits and the trimming of beards and mustaches. Others can't help but notice poor upkeep, even if you don't.

Of course, we all know some very good engineers who are oblivious to such details. You can be sure that their apathy in this regard has been noted by those around them, if not explicitly, certainly subconsciously. Likewise, we all know some "wild" ones. They all must accept whatever the consequence is of their personal appearance, whether they like it or not, and so must you.

Refrain from using profanity in the workplace.

You will be well served if you simply avoid using profane language. Not using profanity will never be offensive to anyone. On the other hand, using profanity may well be offensive, even without your knowing it.

To be sure, obscene and vulgar language is routinely heard in some circles. Unfortunately, in the professional workplace, such language is sometimes used for its "effect." Some think it a mark of power, strength, or vigor. The trouble in using profanity is that its actual effect is known only to the listener, who may conclude something quite different about the speaker from what was intended. Regardless, blatantly obscene language serves no one properly, and a really foul mouth will generally inspire nothing but contempt.

Take it upon yourself to learn what constitutes harassment and discrimination — racial, ethnic, sexual, religious — and tolerate it not at all in yourself, your colleagues, your subordinates, or your company.

There is simply no room in the workplace for harassment or discrimination of any kind. Blatant infractions will get employees and employers into big trouble, as most people know, but big trouble can also come from subtler forms of these unacceptable behaviors. The "harmless" off-color joke and slightly misguided comment can also offend, and are not acceptable.

Of course, no one likes the self-appointed police officer who is constantly on the look-out for rule violations or who constantly cautions colleagues about workplace behavior. Confronting a colleague or subordinate on such matters must itself be handled discretely and delicately. Your best course of action, whether you are a target or an observer of suspected harassment or discrimination, depends upon the circumstances. You may choose to informally approach the alleged offender directly, but any formal discipline should be handled together with your manager, personnel department, or both.

Beware of what you commit to writing and of who will read it.

Be careful about who gets copies of your letters, memos, and messages, in whatever form or medium they are created, especially when the interests of other departments are involved. Engineers have been known to broadcast memoranda containing damaging or embarrassing statements. Of course it is sometimes difficult, especially for novices, to recognize the "dynamite" in such a document. But, in general, it is apt to cause trouble if it steps too heavily into another's domain or reveals serious shortcomings on anybody's part. If it is to be distributed widely or if it concerns manufacturing or customer difficulties, you'd better get a higher authority to review and approve it before it goes out.

Once you have issued something in writing, despite your best attempts to the contrary, you will have relinquished control of its distribution and its life. To be safe, you had better assume that your documents might go to anyone and that they will exist forever. Compose them accordingly.

As the worst conduct in this regard, misplaced verbal assaults can cause enough mischief, but putting such emotions into writing can cause way more than enough. Anger, malice, disrespect, and ridicule toward another will be remembered in written documents forever, which may be long after you wish they had been forgotten.

Beware of using your employer's resources for personal purposes. It may be considered suspicious at best, and larcenous at worst.

Most of us have used the office copier or borrowed a shop tool for our personal use, and we trust that hardly anyone would ever make anything of it. Nevertheless, whenever you use your company's property, equipment, or supplies for anything other than company business, you risk suspicion. Furthermore, your employer has every right to investigate your behavior, including examining all of your "personal" domain at work (for it is not really personal) for any evidence of misconduct.

To be sure, most employees, absent obvious transgressions in this regard, should be held above suspicion, and the first examples above are probably harmless enough. Nevertheless, for what little you will likely gain by appropriating anything of your employer's assets, in the end it hardly seems worth the risk, even if only to your integrity.

REGARDING CAREER AND PERSONAL DEVELOPMENT

Maintain your employability as well as that of your subordinates.

It is the rare engineer who has a single employer for a whole career, and employers understand this. So it follows that it is unreasonable to expect engineers to accept becoming useless to other potential employers, however invaluable they may have become to their current employer. If your skills and knowledge are valuable only to your current employer, you are in trouble. Sooner or later, for one reason or another, your employer will no longer be interested in buying those skills, and you will have no place else to sell them.

Obsolescence is bad business for employers as well as employees. It is costly for employers to disposition the obsolete, and to hire or develop employees with the skills that the departed should have been developing all along. Therefore, for the benefit of your employer, you should also make this situation unequivocally clear to your subordinates, then you should do all you can to counsel and support them in this regard.

Be an adherent and a proponent of life-long learning. This doesn't mean constant formal training — university classes, seminars, short courses, company-sponsored training — although some of these are a necessary part of a life-long employability plan. It also means taking more than a passing interest in your field by reading sales literature, trade magazines, and professional publications, and attending trade shows and professional conferences. You must find ways to keep up-to-date on new technology in your field regardless of how much or how little your employer supports you, and if you wait for someone to offer you the opportunity, you are just waiting for your own obsolescence.

All of this will require sacrificing some personal time and, perhaps, some personal expense as well. Simply put, not all employers will accept the full burden of employees' continuing education. The effort and dedication required to remain employable is in every engineer's best interest.

Analyze yourself and your subordinates.

Engineers and engineering managers need not be students of psychology — most are disinclined anyway. Nevertheless, it is enlightening to appreciate that people, including yourself, behave as they do not so much because they want to behave that way, but rather because that is how they are. Fundamentally, people see and react to things, and judge and decide things quite differently from one another. Even without fully understanding different personality types, simply recognizing that people are remarkably different will help you accept different personalities as normal, and not to view them as somehow wrong.

Among the important decisions for engineers to make for themselves or their subordinates are when and how much managerial and administrative responsibility is appropriate. It has been assumed in all the foregoing that any normal individual will be interested in either: (a) advancement to a position of greater responsibility, or (b) improvement in personal effectiveness as regards quantity and quality of accomplishment. Either of these should result in increased financial compensation and satisfaction derived from the job. As to item (a), it is all too often taken for granted that increased executive and administrative responsibility is a desirable and appropriate form of reward for outstanding proficiency in any type of work. There should be other ways of rewarding an employee for outstanding accomplishment, for such rewards may be a mistake from either of two points of view:

1. People are sometimes surprised to find that they are much less happy in a new, higher-level job than they thought they were going to be. It is not uncommon for engineers or scientists to discover, to their dismay, that as soon as they become managers they no longer have time to be engineers or scientists.

2. It by no means follows that a good engineer will make a good manager. Many top-notch technologists have been promoted to administrative positions much to their own and their employer's detriment.

These facts should therefore be considered carefully by the person threatened with promotion and by the person about to do the promoting. It is not always easy, however, to decide in advance whether you, or the employee in question, would be happier and more effective as more of a manager and less of an individual worker, or vice versa. There is no infallible criterion for this purpose but it will be found that, in general, the two types are distinguished by the characteristics and qualities listed in the accompanying table.

Although nobody fits neatly into one column or the other, a preponderance of engineers has the qualities of the specialist column. This seems to be especially true for engineers just starting their careers; most will have been drawn to engineering because they have these characteristics. This, no doubt, is the source of the conventional stereotype of the engineer. But beware — you do a grave disservice to those who you assume fit this stereotype simply because they are engineers. Everyone is a unique blend. Furthermore, none of these characteristics should be considered better or worse than its counterpart, but each is quite clearly different from the other.

In reality, nobody successfully moving through an engineering career can avoid management and administration altogether.

CHARACTERISTIC QUALITIES
OF MANAGERS OR INDIVIDUAL SPECIALISTS

Manager	Individual Specialist
Extrovert	Introvert
Cordial, affable	Reserved
Gregarious, sociable	Prefers own company
Likes people	Likes technical work
Dominant	Unassuming
Seeks to understand other views	Seeks the facts of the matter
Likes organizing things	Likes doing things
Interested in:	Interested in:
• business	• science and mathematics
• costs	• devices
• use	• function
• practices	• principles
Ability to get many things done	Ability to get intricate things done
Excels at communication	Excels at analysis
Fast, intuitive decision making	Methodical decision making
Talent for leadership	Independent, self-sufficient
Impulsive	Intellectual
Vigorous, energetic	Meditative, philosophical

These are necessary parts of all job descriptions, and a certain amount of managing projects and supervising others is satisfying for all but the most narrow-minded technologist. Further, as time goes by, many engineers find their interest in management changing, often increasing as their career matures.

Although certain personality types are more disposed to become managers in their careers, and some personality characteristics seem to be more common than others in successful managers, any personality type can be that of a successful manager, and no personality characteristic precludes someone from managerial success. We all know the reserved, introspective, or intellectual, yet highly effective manager. What makes a successful manager, and whether or not a person will thrive as a manager, is more complicated than simply matching up a few traits. Indeed, those who choose to become managers can become successful regardless of their traits by manipulating their situations and selecting their styles to make the best use of strengths and to de-emphasize weaknesses. Perhaps as important as anything, upon becoming, or in anticipation of becoming, a newly promoted manager, anyone is well advised to seek the knowledge and training required to succeed. Time and again it is all too clear how seriously ill-equipped many engineers, with their technical training alone, are to be managers.

As regards analyzing yourself and your subordinates, some final good advice for anyone is: Do what you do best; you will also be the happiest. Try to improve your best traits and make them the most visible. Of course, try to improve anything of yourself that people could consider substandard, but trying to become expert, or even proficient at something for which you have little natural talent is futile. It is often better to minimize the harm by making it unnecessary in your job or invisible in your behavior. Analyze yourself and then emphasize the positive!

CONCLUSION

The foregoing "laws" represent only one element in the general formula for a successful engineering career. However much natural interest you take in these principles — and each person has his or her own level — it will pay for you to contemplate at least a little of the "rules of the game" to develop your own set of principles and practices to guide you through your professional career. To engineers interested in improving their professional effectiveness further, study in not only technical, but also in professional and administrative subjects is recommended; there is no shortage of published material to be found in bookstores and libraries. Finally, it should be observed that the various principles that have been presented, just like those of science and engineering, must be assiduously applied and developed in practice if they are to become effective assets.

REPRESENTATIVE
PROFESSIONAL
CODES OF ETHICS

SOCIETY POLICY
ETHICS

ASME requires ethical practice by each of its members and has adopted the following Code of Ethics of Engineers as referenced in the ASME Constitution, Article C2.1.1.

Code of Ethics of Engineers

The Fundamental Principles

Engineers uphold and advance the integrity, honor and dignity of the engineering profession by:

I. Using their knowledge and skill for the enhancement of human welfare;

II. Being honest and impartial, and serving with fidelity the public, their employers and clients; and

III. Striving to increase the competence and prestige of the engineering profession.

The Fundamental Canons

1. Engineers shall hold paramount the safety, health and welfare of the public in the performance of their professional duties.

2. Engineers shall perform services only in the areas of their competence.

3. Engineers shall continue their professional development throughout their careers and shall provide opportunities for the professional and ethical development of those engineers under their supervision.

4. Engineers shall act in professional matters for each employer or client as faithful agents or trustees, and shall avoid conflicts of interest or the appearance of conflicts of interest.

5. Engineers shall build their professional reputation on the merit of their services and shall not compete unfairly with others.

6. Engineers shall associate only with reputable persons or organizations.

7. Engineers shall issue public statements only in an objective and truthful manner.
8. Engineers shall consider environmental impact in the performance of their professional duties.
9. Engineers shall consider sustainable development in the performance of their professional duties.

The Board on Professional Practice and Ethics maintains an archive of interpretations to the ASME Code of Ethics (P-15.7). These interpretations shall serve as guidance to the user of the ASME Code of Ethics and are available on the Board's website or upon request.

Reponsibility: Council on member Affairs/Board of Professional Practice and Ethics

Adopted: March 7, 1976

Revised: December 9, 1976
               December 7, 1979
               November 19, 1982
               June 15, 1984
               (editorial changes 7/84)
               June 16, 1988
               September 12, 1991
               September 11, 1994
               June 10, 1998
               September 21, 2002
               September 13, 2003

THE INSTITUTE OF ELECTRICAL AND ELECTRONICS ENGINEERS (IEEE)
Code of Ethics

We, the members of the IEEE, in recognition of the importance of our technologies affecting the quality of life throughout the world, and in accepting a personal obligation to our profession, its members and the communities we serve, do hereby commit ourselves to the highest ethical and professional conduct and agree:

1. to accept responsibility in making engineering decisions consistent with the safety, health and welfare of the public, and to disclose promptly factors that might endanger the public or the environment;
2. to avoid real or perceived conflicts of interest whenever possible, and to disclose them to affected parties when they do exist;
3. to be honest and realistic in stating claims or estimates based on available data;
4. to reject bribery in all its forms;
5. to improve the understanding of technology, its appropriate application, and potential consequences;
6. to maintain and improve our technical competence and to undertake technological tasks for others only if qualified by training or experience, or after full disclosure of pertinent limitations;
7. to seek, accept, and offer honest criticism of technical work, to acknowledge and correct errors, and to credit properly the contributions of others;
8. to treat fairly all persons regardless of such factors as race, religion, gender, disability, age, or national origin;
9. to avoid injuring others, their property, reputation, or employment by false or malicious action;
10. to assist colleagues and co-workers in their professional development and to support them in following this code of ethics.

Approved by the IEEE Board of Directors, August 1990
Internet Reference:
www.ieeeusa.org/documents/career/career_library/ethics.html

THE AMERICAN SOCIETY OF CIVIL ENGINEERS (ASCE)

Code of Ethics

Fundamental Principles

Engineers uphold and advance the integrity, honor and dignity of the engineering profession by:

1. using their knowledge and skill for the enhancement of human welfare and the environment;
2. being honest and impartial and serving with fidelity the public, their employers and clients;
3. striving to increase the competence and prestige of the engineering profession; and
4. supporting the professional and technical societies of their disciplines.

Fundamental Canons

1. Engineers shall hold paramount the safety, health and welfare of the public and shall strive to comply with the principles of sustainable development in the performance of their professional duties.
2. Engineers shall perform services only in areas of their competence.
3. Engineers shall issue public statements only in an objective and truthful manner.
4. Engineers shall act in professional matters for each employer or client as faithful agents or trustees, and shall avoid conflicts of interest.
5. Engineers shall build their professional reputation on the merit of their services and shall not compete unfairly with others.
6. Engineers shall act in such a manner as to uphold and enhance the honor, integrity, and dignity of the engineering profession.
7. Engineers shall continue their professional development throughout their careers, and shall provide opportunities for the professional development of those engineers under their supervision.

As adopted September 25, 1976 and amended October 25, 1980, April 17, 1993 and November 10, 1996.

Further guidelines on interpretation are available at:
http://www.asce.org/inside/codeofethics.cfm

BIBLIOGRAPHY

NOTE: The 1944 publication of this work included many references in addition to these, including both books and published papers. Most have been omitted from this edition for being either unavailable or no longer applicable. Citations for specific text from the original have been retained.

Adams, James L. *Conceptual Blockbusting*, Perseus Press, 1990. ISBN 0201550865.

Adams, James L. *Flying Buttresses, Entropy, and O-rings, The World of an Engineer*, Harvard University Press, 1991. ISBN 0-674-30688-0 (Paperback ISBN 0-674-30689-9).

Badawy, Michael K. *Developing Managerial Skills in Engineers and Scientists: Succeeding as a Technical Manager*, John Wiley & Sons, 2nd Edition, 1995. ISBN: 0-471-28634-6.

Baldwin, Lionel V., and Marvin F. DeVries. "Take Care of Yourself: Stay Employable." *Manufacturing Review*, **8**, 1 (March 1995), 78–85.

Baron, Renee. *What Type Am I? Discover Who You Really Are*, Penguin Books, 1998. ISBN 0-965-68489-X.

Borchardt, John K. *Career Management for Scientists and Engineers*, American Chemical Society, 2000. ISBN: 0841235252.

Buckingham, Marcus, and Curt Coffman. *First, Break All the Rules*, Simon and Schuster, 1999. ISBN 0-684-85286-1.

Ferguson, Eugene S. *Engineering and the Mind's Eye*, The MIT Press, 1992. ISBN 0-262-06147-3.

Fleddermann, Charles B. *Engineering Ethics*, Prentice-Hall, 1999. ISBN 0-13-784224-4.

Florman, Samuel C. *The Civilized Engineer*, St. Martin's Griffin, 1987. ISBN 0-312-00114-2 (Paperback ISBN 0-312-02559-9).

Florman, Samuel C. *The Existential Pleasures of Engineering*, St. Martin's Griffin, Second Edition, 1994. ISBN 0-312-14104-1.

Florman, Samuel C. *The Introspective Engineer*, St. Martin's Griffin, 1997. ISBN 0-312-15152-7.

Glassman, Audrey. *Can I FAX a Thank-You Note?*, Berkeley Books, New York, 1998. ISBN 0-425-16433-0.

Goleman, Daniel. *Working With Emotional Intelligence*, Bantam Books, 1998. ISBN 0-553-37858-9.

Gough, H. G. "The California Psychological Inventory." Chapter in *Major Psychological Assessment Instruments, Volume II*, Charles S. Newmark, Ed., Allyn and Bacon, 1989.

Gough, Harrison G. "A Managerial Potential Scale for the California Psychological Inventory," *Journal of Applied Psychology*, **69**, 2, 1984, 233–240.

Griffin, Jack. *The Unofficial Guide to Climbing the Corporate Ladder*, IDG Books Worldwide, Inc. 1999. ISBN 0-02-863493-4.

Jacobs, Bruce A. *Race Manners: Navigating the Minefield Between Black and White Americans*, Arcade Publishing, 1999. ISBN: 1559704535.

Jain, R. K., and H. C. Triandis. *Management of Research and Development Organizations, Managing the Unmanageable*, John Wiley & Sons, second edition, 1997. ISBN 0471146137.

Katz, Ralph (Editor). *The Human Side of Managing Technological Innovation, A Collection of Readings*, Oxford University Press, New York, New York, 1997. ISBN 0-19-509693-2 (Paperback ISBN 0-19-509694-0).

Keirsey, David, and Marilyn Bates. *Please Understand Me, Character & Temperament Types*, Prometheus Nemesis Book Company, 1984. ISBN 0-9606954-0-0.

Keirsey, David. *Please Understand Me II: Temperament, Character, Intelligence*, Prometheus Nemesis Book Co., 1998. ISBN 1885705026.

Layton, Edwin T., Jr. *The Revolt of the Engineers: Social Responsibility and the American Engineering Profession*, Case Western Reserve Press, 1971. ISBN 0801832861.

McAllister, Loring W. *A Practical Guide to CPI Interpretation*, Consulting Psychologists Press, 1996.

McGrath, Joseph. E. *Groups: Interaction and Performance*, Prentice-Hall, Englewood Cliffs, NJ, 1984. ISBN 0-13-365700-0.

Myers, Isabel Briggs, and Peter B. Myers. *Gifts Differing, Understanding Personality Type*, Consulting Psychologists Press, 1980. ISBN 0-89106-064-2.

Osborne, H. S. "Definition of an Executive," *Electrical Engineering*, **61**, August 1942, p. 429.

Peddy, Shirley. The Art of Mentoring : *Lead, Follow and Get Out of the Way*, Learning Connections, January, 1999. ISBN: 0965137635.

Pinkus, Rosa Lynn B., Larry J. Shuman, and Norman P. Hummon. *Engineering Ethics: Balancing Cost, Schedule, and Risk: Lessons Learned From the Space Shuttle*, Cambridge University Press, 1997. ISBN 0-521-43171-9.

Rabbe, Willis. "Administrative Organization for a Small Manufacturing Firm." *Mechanical Engineering*, **63**, 1941, 517–520.

Project Management Institute. *A Guide to the Project Management Body of Knowledge*, 1996 edition, PMI Publishing Division, Sylva, N.C., 1996. ISBN 1-880410-13-3 (Paperback ISBN 1-880410-12-5). (www.pmi.org)

Schell, Erwin Haskell. The *Technique of Executive Action*, fifth edition, McGraw-Hill Book Company, Inc., New York, 1942.

Smith, E. D. *Psychology for Executives*, Harper & Brothers, New York, 1935.

Vargo, John F. "Professionally Speaking: Understanding Product Liability." *Mechanical Engineering*, **117**, 10, 1995, 46.